THE SOLAR SYSTEM
BUILT BY AN ENGINEER

JAMES L. BAKER
AEROSPACE ENGINEER RETIRED

ISBN: 1453764658
ISBN-13: 9781453764657
Library of Congress Control Number: 2010912222

Cover Photo credit: X-ray: NASA/CSC

CONTENTS

PREFACE

I wrote this book to document a new and revolutionary theory for the formation of the Solar System that is totally at odds with the standard model and any other model. I believe I have a fundamentally sound theory that is consistent with the laws of physics and produces exactly what is here. The first premise of the theory is that the planets or at least all of the planetesimals, formed before the Sun formed. As odd as this may sound, the conclusion is based purely on the laws of orbital mechanics. The second major premise is that all liquid and solid materials in the Solar System are products of combustion. Chemical reaction is the only source of energy that was sufficient to produce planets with the characteristics the planets have.

INTRODUCTION

When anyone looks at the sky and sees a huge, magnificent object like the Sun, it is only natural to assume the Sun is the parent of its domain. The Sun has always been worshiped as a god and the planets are like children. Why would anyone think otherwise? It has always been assumed that the Sun came first. Scientists starting from the assumption that the Sun formed first have come up with many ways to build a star. Generally they have a dust cloud collapse like a self eating watermelon and then the nuclear furnace fires up. However, they run into real problems when they try to explain how the planets formed. Building planets is a whole different ball game. Almost every month, it seems that the astronomy media comes up with a new planet formation theory. None of these theories deliver the finished products. They would be lucky if they produced a dust ball. Planets like ours must end up in near circular orbits around the star, rotate, have molten centers, and a planet like earth must end up with all sorts of special features.

I am proposing a more natural, slow process for building the Solar System. The task starts with a nebula made up of solids and gases. The materials first form into small masses (planetesimals) and then all but about 0.1% of the mass collapses into the center to form the Sun. While the Sun is building, planetesimals are merging far out from the center and begin the process of merging into planets. The growing planets are attracted by the Sun and slowly spiral into their final orbits.

The first point that must be made is to show that the Solar System planets could not have formed after the Sun. If the Sun had formed first, the planets would rotate about their axes in the opposite direction from the way they actually rotate. It must be concluded that the planets did not form after the Sun.

If the planets form before the Sun, they must form far out in the nebula or they would be captured by the Sun. This conclusion establishes the baseline for all that follows.

Next, it seems odd that no one ever seemed to give any consideration to the gases, metals, and other elements when explaining the formation of planets. Oxygen is the third largest constituent in the makeup of the Solar System, and yet it isn't even acknowledged in any of the theories I've read. Apparently this was the result of building the Sun first. If the Sun came first, it would collect all of the gases anywhere near it and all that would be left in orbit is some dust. Surely oxidation of metal dust in a near vacuum didn't produce all of the sand, salt, minerals, and water and then become dust and crash into each other. This just isn't a good way to build something.

There is a much more straightforward method for building the Solar System by just using gravity and chemistry.

The approach that I used to develop this new theory was to look at the makeup and behavior of the end products in the Solar System and determine known processes that could have created them. As I worked through time, the processes produced a sequence of logical events that explained:

Why the planets spin and orbit counterclockwise

Where the planets were formed

Why the planets are in their particular orbits

How planet spin was generated

How the planets attained their high spin rates

Why the core of the earth is molten

How the heavy metals happen to be in the center of the earth

What was the source of our atmosphere

Why all planets should have or had water, carbon dioxide and Nitrogen in their atmospheres or frozen on their surfaces

How the Sun was formed

Why the angular momentum of the Sun is not proportional to its share of the Solar System's mass

This was all accomplished using fluid dynamics, Newtonian physics, basic chemistry and a little thermodynamics.

Hopefully this theory presents some new ideas that explain the current state of the Solar System better than other theories. After searching the internet and reading some of the theories on the formation of the Solar System, I have not come across a more straightforward scenario. The processes in this paper are original thoughts of the author although they may not be unique. Here is the story.

1. THE SOLAR NEBULA.

We begin this theory where the material that formed the Solar System was produced by a supernova and was separated from the mother cloud. The material had cooled to temperatures where gases existed as molecules. The nebula was in the cold, dark of space and had radiated to a temperature near absolute zero. For reasons explained later, it is assumed that the elements were produced in waves or clouds and did not mix into a homogeneous brew. The initial Solar Nebula is assumed to be more spherical than flattened since that shape is more likely to produce a star in the center that has over 99.9% of the Solar System's mass. The nebula had to have been rotating around an axis in order for the current Solar System to be rotating counterclockwise. The direction of rotation is defined as the view from Polaris. Since the initial Solar Nebula would not have had a concentrated mass at the center, the circular velocity of any mass was a direct function of its distance from the axis of rotation. Think of a rotating solid ball. Gravity and centrifugal force established the shape of the nebula. It is safe to assume that the angular velocity of the material out where Pluto is now was not much above or below the velocity that Pluto has now. A rotating cloud of material must be held together by gravitational forces countering centrifugal forces or the outer material would escape. However, a cloud of material, in any shape, in space, cannot reach steady state until nearly all of the mass is in the ecliptic plane. Initially, matter above and below the ecliptic plane will be pulled toward the ecliptic plane and toward the center of rotation. Matter that does not have sufficient centrifugal force to resist the ever increasing gravitational pull building in the center would be pulled into the forming Sun, eventually. As the mass in the center increases, the outer edge of the ecliptic plane will be drawn inward and the angular velocity of material there will accelerate to balance the pull of the more concentrated mass in the center. Any mass which does not end up in the center has to be in orbit. That is the state of the current Solar System.

The gravitational pull on and therefore the acceleration of any matter in the initial cloud would have been very low considering that the mass was distributed throughout a cloud that was probably more than 100 AU* across. It is certain that the gravitational force of the nebula anywhere in the nebula could not be greater than it would be at that same distance with the Sun in place. To put the density of the nebula into perspective, envision a BB size Sun in the middle of the 5th floor of a 10 story building. Now consider that all of the mass originally in the building is that which created the BB. The whole process of collapsing the nebula must have taken billions of years.

*AU (Astronomical Unit) The mean distance from the Sun to Earth ~ 93 million miles.

2. HOW PLANETS CAN NOT BE FORMED.

The first point that I want to make is the foundation for this whole theory, and that is that **the planets in the Solar System had to have begun forming before the Sun formed.** If any type of material is orbiting around or is spiraling toward a massive center such as the Sun, you will have a condition somewhat similar to a hurricane or water going down a drain. In the Solar System this revolving is counterclockwise as viewed from above the Sun's North Pole. The laws of orbital mechanics require that orbiting masses closest to the Sun have the highest orbital velocities and that velocities decrease as distance from the Sun increases. For example, Mercury revolves at 107,172 mph, Earth at 66,615 mph, while Jupiter revolves at 29,194 mph. Any bounded volume of material you consider that is revolving around the Sun must have a net rotation in the clockwise direction. Picture a circle revolving around a point with the outer edge going slower than the inner edge. The circle will elongate and the outer edge will lag behind. If this were the case during planet accretion, material impacting an accreting planet from the Sun-side would have more momentum than material coming in from the other side. Due to the angles of the materials' velocity vectors at impact, the resulting momentum exchange would cause a clockwise torque on the building planet. You would end up with a clockwise spinning planet. We would have the Sun rising in the west and setting in the east if this were the case.

The explanation I usually hear for everything rotating counterclockwise in the Solar System is that it is due to conservation of angular momentum of the Nebula. These explanations are almost always accompanied by the analogy of a skater pulling in her arms which causes her to spin faster. The angular momentum of a cloud of material revolving around the Sun once a year would be very, very low. It is like a skater going around the rink once a year and pulling in her arms. Any material in a cloud revolving around the Sun would not have anywhere near enough angular momentum to produce a rotation rate like Earth's. As a matter of fact, the low level of the angular momentum of the nebula is one of the strongest arguments for the planets-first theory. Analysis shows that it would take the merger of 20 planetesimals rotating 7 times per Earth day to produce the angular momentum of the Earth. This is assuming totally efficient merging which of course it wouldn't be.

Something that seems to be overlooked by the dust collapse theories is that conservation of angular momentum only applies if it is not interfered with. Angular momentum can easily be converted to other forms of energy. Brakes convert it to heat. It can be used to make electricity or pressurize gas. It can be used to crush rocks. If the twirling skater's skates dig in that's the end of conserved angular momentum. In other words, angular momentum is not like the speed of light.

As a fluid flow analyst, it appeared to me that the torque required to cause the high rotation rate of a planet had to come from more than the collapse of a cloud of material within the nebula. A cloud within the initial nebula would not have any appreciable rotation. The only source of rapid rotation had to involve the gases that were present throughout the nebula.

3. WHEN PLANETS FORMED.

Since we know the planets do rotate counterclockwise, it must be concluded that they did not form in orbit around the Sun. **The planetesimals formed before the Sun formed.** They formed in the counterclockwise rotating Solar Nebula where the material with the highest velocity was on the outer edge of the ecliptic plane. Velocity distribution in a rotating nebula is equivalent to that in a solid ball. Now, material accreting from the outside of a building planet has a higher velocity than the material closest to the center of the nebula, thus counterclockwise rotation of the planet.

I would like to point out that there is nothing we observe in everyday life similar to the phenomenon of a mass accreting in the flow field around a vortex center. The nearest comparison might be the formation of a tornado in a hurricane. This lack of a familiar occurrence may be why no one has given serious thought to the direction of rotation of the planets.

4. SOURCE OF PLANET MATERIAL.

After calculating a possible average density of the Solar Nebula by dividing the mass of the Sun by the volume of a 100AU diameter sphere, I obtained a value of 475 kilograms per cubic mile or 71 trillionths of a pound per cubic foot. The notion that dust of that density could collapse into spinning planets doesn't seem reasonable. A more likely scenario was to have high density clouds made up of mostly single elements distributed throughout the nebula. There are no logical processes by which Earth's iron oxide, silicon dioxide, gold nuggets, etc. could have developed due to gravity or temperature if the elements were all evenly mixed in a cloud. Also, if all of the elements were mixed initially as atoms in the nebula, the hydrogen and oxygen would have combined and only water would exist as an oxide in the universe. Having clouds of specific elements scattered throughout the nebula required that the elements were generated in waves or bubbles by the Supernova. This must be the way supernovae feed out atoms. The discovery of G292.0+1.8 by NASA's Chandra X-ray Observatory tends to corroborate this scenario. The illustration below shows the separate concentrations of the elements.

G292.0+1.8

5. WHERE THE EARTH FORMED.

Since the planets formed before the Sun, then the existing planets had to have formed far out on the edge of the ecliptic plane or the Sun would have captured them as it was acquiring over 99.9% of the mass in the Solar System. As the assembly of the Sun proceeded, the matter in the outer reaches of the disk was in an ever decreasing toroidal volume. The section view of the toroid would be more tear drop shaped with the small end toward the Sun. That is the part of the nebula where the existing planets had already formed. Those planets were very far from the Sun when they were formed and experienced ever increasing pull toward the building Sun as it increased in mass. The mass that became Mercury formed the inner boundary of the mass that escaped. Mercury would have had the slowest angular velocity and shortest distance from the Sun when it formed and therefore would end up the closest to the Sun. The final orbital velocities of the planets are the result of the work done on them by the Sun's gravity. Their initial circular velocities in the nebula were relatively slow.

6. WHY CIRCULAR ORBITS.

The near circular orbits of the existing planets provides another indication that the planets came in from far out in the Solar System. Only by starting their journey from a circular orbit where the building Sun's gravitation was just slightly greater than the centrifugal force could they finally reach a balance of the two forces when they reached their final orbit. Very little initial velocity of a body toward the Sun has a large effect on the shape of the final orbit that is reached. A large entry angle will cause the body to end up in the Sun or on a highly elliptical orbit.

 As stated above, if the planets formed in place or from dust after the Sun formed, they would have formed under orbital conditions and rotated clockwise.

The large gap between Mars and Jupiter which you would think should have been populated with planets is not easy to explain. I can only surmise that Jupiter, because of its great size captured what was in that space. Jupiter and the other planets began forming so far from the center of the nebula that they formed much like the Sun. Planetesimals along with their hydrogen and helium accreted to form these planets.

The big rocks in the Asteroid Belt are most likely debris from solid planetesimals colliding at high speeds. They are too small and far apart to accrete into anything large.

7. PRODUCTS OF COMBUSTION THEORY.

This part of the theory of the formation of the earth and the other planets deviates substantially from any other theory I have found. The theory poses a new process for creating the planets. **The planets are products of combustion.** That is to say, they were cooked, not pressed.

Some theorize that Earth's molten core is due to pressure, some say it is due to kinetic energy from collapsing dust, asteroids or planets colliding at hyper velocities and some say it is due to nuclear fission. None of these explanations quite seem to fit since they lead to a homogeneous planet with little rotation and no atmosphere. None of these processes do a good job of explaining where the water, mineral diversity and our atmosphere came from. Gravity is not a very good separator of heavy molecules from lighter ones in a fluid mixture either. The closer you get to the center of a planet, the lower the gravitational force. Likewise, volcanoes are not good builders of atmospheres or oceans. They just don't come up with the right combination and quantities of materials.

After many years of working with cryogenic rocket propellants and observing them in the gas, liquid and solid states, I have developed a different view of the sequence of events that would occur in cold, dark space. The heavier elements such as iron, magnesium and silicon were likely in mostly single element clouds throughout the nebula somewhat like vapor clouds in our sky. The dust and metals in the clouds were intermixed with the gases that had spread throughout the nebula, mainly hydrogen, helium, oxygen, nitrogen, and chlorine. These clouds could have been any size, up to several AU across. Without having clouds of higher mass densities, the chances of building planets are low. Each of the clouds would slowly collapse into counterclockwise rotating disks of the metal dust and gases. These clouds would have been everywhere throughout the nebula. Based upon the composition of the solar system, oxygen, next to hydrogen, was the most prevalent reactive gas in the cloud. At the near absolute zero temperatures in space, these gas molecules would be in the solid state. All gases except possibly hydrogen and helium would have attached to the solids as ice. (This would be similar to the normal occurrence of cryopumping of air onto liquid hydrogen ducts on spacecraft.) The origins of planetesimals were therefore disks of material made up of a mixture of dust, carbon, metals and solidified gases. As anyone who has witnessed a pure oxygen fire knows, Oxygen does not like to coexist with anything that will burn. At some level of a planetesimal's gradual buildup, an impact or electrical spark would ignite this mixture. The heat of combustion would have melted all dust and small rocks which, along with the metal oxides and water, eventually would coalesce due to gravity. (This process of combustion and melting could be readily evaluated in a cryogenic test facility.) The gaseous products of combustion of the oxygen with carbon and hydrogen (CO_2 and steam), ultimately would be displaced by the melted solids and the gases would move to the surface of the melted material along with the Nitrogen and any leftover Hydrogen. The steam, hydrogen, CO_2, and nitrogen in the atmosphere would then expand and cool, radiate to space, condense to liquid or solid, fall back and eventually cool the outer surface of the melted solids causing the outer layer of the melted material to solidify. Eventually the water and other gases would solidify in layers. This process would continue until the cloud was accreted into a planetesimal. Once a piece of solid material developed, the accretion process would be off and running. The core of the planetesimal was a melted mineral. Some would be mostly iron oxide, some Silicon dioxide or magnesium oxide and some would be mixtures. The process of merging planetesimals into planets is described below.

8. SOURCE OF PLANET ROTATIONS.

We know that the collapse of a relatively dense cloud of matter rotating around the nebula axis will cause some slight counterclockwise rotation of a forming planetesimal. The cloud was probably revolving once every 250 years or so. As additional mass comes into the rotating core, the work energy added to the incoming mass by gravity would increase the rotation rate of the disk. This effect was a major factor in creating the initial rotation rates of the planetesimals. However, the small quantities of angular momentum from the clouds collapses are nowhere near sufficient to add up to the angular momentum that the Earth has. As stated above, we need twenty planetesimals that have 1/20 of the earths mass, rotating more than 7 times a day, to merge and produce Earth's rotational angular momentum.

Identifying the source of rapid planet rotation was not an easy problem to solve. I wish I could have come up with an easier process to accomplish the planet spin up, but I can't. The only source of such a large amount of energy needed to produce the Earth's angular momentum has to be from chemical reaction. But how could that produce torque? The answer is that the conflagrations that generated the planetesimal core material provided the energy. The heat of combustion caused the core solids and gases to expand outward, probably a great distance. Remember that this material had angular rotation already from the initial collapse. Gravity would eventually cause the outward bound material to collapse again, along with any new material that was accreted. It is difficult to give an analogy of the effect of a conflagration of material in space. There is no way physically to simulate the effects of gravity and near vacuum simultaneously. When a skater pulls in her arms, she does work and she and her arms speed up. When she quits restraining her arms they move outward and do work on her and they both slow down. In the forming planetesimal, the heat of combustion causes the gases to expand and move outward along with some solids. This has no effect on the angular velocity of any of the mass. There is no tangential force applied by the expansion. The force of gravity will be toward the center of mass. After gravity stops the outward motion, the expanded material moves back in. Conservation of momentum will pull the material back in resulting in a higher angular velocity. The solid core rotation will not be affected until the returning material impacts it. The net effect of this event is that the entire mass will rotate faster. If this process were repeated several times, the material would end up with a very high rotation rate indeed. Pressure of the heated gases did the work of pushing out the mass and gravity pulled the mass back in. We now have rapidly spinning planetesimals and can proceed to build rotating planets.

Since there is almost no drag in space on the rotating solid mass, the angular velocity will not decrease over time.

9. PLANETESIMAL ACCRETION.

Planetesimals would begin merging throughout the nebula as soon as they attained sufficient gravity. Initially, they would be rotating around the nebula's axis and their angular velocities would be directly proportional to their distance from the axis. There is no reason to assume that planet formations were not happening throughout the nebula. The Sun has a mass of 330,000 Earths so millions of planetesimals would have formed. As mass increased in the center of the nebula, this mass would attract more and more planets and planetesimals. The Sun would begin slowly forming. Eventually, nearly all mass above and below the ecliptic plane would be pulled into the Sun at an ever increasing rate. Planetesimals above and below the ecliptic plane that were far out from the center would be pulled by their mutual attractions into paths near the ecliptic plane where they would merge. As soon as the Sun's gravity was sufficient, all of the material in the ecliptic plane would begin to spiral very slowly toward the Sun. It would be illogical to conclude that a complete planet formed in its orbit on one side of the Sun. As planetesimals spiraled closer to the building Sun, the ones starting from the closest locations would have had the lowest centrifugal force and would attain higher final orbital velocities than the others due to the work energy added by the Sun's gravity. As planetesimals in a band around the ecliptic plane spiraled in toward the Sun, their paths would cross and they would meet unless they were exactly the same distance from the Sun. I would like to point out that the closing speed between two small masses, which is attained due to gravitational attraction, is low. We are used to thinking of speeds that are caused by the attraction of the sun and planets, but that would not be the case for dust and planetesimals. The escape-velocity/incoming-velocity for small planetesimals could be in the low m/sec range.

As far as tilted orbital planes are concerned, these conditions would just be the result of a bias of accreted mass from above or below the ecliptic plane

10. PLANETESIMAL MERGING

The merging process for planetesimals spiraling inward toward the building Sun is bound to be complex. How two planetesimals travelling at about the same velocity in about the same orbit would attract each other would depend on their total sizes and relative sizes. Smaller masses would approach each other slowly. A large mass would pull in a smaller one at a higher speed. If both planetesimals are mostly molten material inside a solid shell and surrounded by gases, they could join like two balls of clay. If both are rotating, the merging becomes even more complex. It is interesting to envision what occurs when two counterclockwise rapidly rotating balls of liquid materials merge at relatively low velocities. They may form a rotating couple and then merge into a spherical ball, or one may move above the other and form a new larger ball. In either case, the material of the ball made up of the lowest density mineral will eventually surround the heavier ball. Magnesium oxide and silicon dioxide would surround iron oxide. If a ball with an inner and outer layer merged with another, the final results would be the same and the heavier material in the center would continue to grow. If a small ball of a particular mineral impacted a large one, it might just spread out wherever it hit. This might explain why such metals as gold and uranium oxide are only found in certain locations. One thing that is certain is that the solidified gases and hydrogen gas around any of the merging bodies will eventually surround the combined final body. The outer surface will cool and solidify since there is not yet any solar radiation.

As a result of all of this construction and relocation, we have finally assembled an earth with a 35 mile thick layer of dust and rocks sitting on a mantle of liquefied rock. The crust will have a very uneven distribution of chemical compounds in it due to the varied makeup of the planetesimals that went into the formation. The Earth is in its orbit around the hot Sun. Most of the ice has melted and we have millions of cubic miles of water.

Ultimately, as a result of such merging, Mercury, Venus, Earth, Mars and the outer planets and asteroids are what remained. All planets surfaces will have water, CO_2, and N_2. If they are far from the Sun, they will also have held onto their H_2 and He. It is no surprise to me when astronomers find water on Venus, Mars and the Moon. That is how they were made.

11. MOONS

The only explanation I can come up with for the existence of moons is that they came to the planets in the same way planets came to the Sun. A large planetesimal attracted a smaller one from a distance that put it in orbit instead of colliding with it. This is easier to envision for the large outer planets than it is for the Earth. Once the Sun has formed, any moon picked up after that time should have a retrograde orbit.

12. SOURCE OF ATMOSPHERES.

By running the numbers, it turns out that the heat of combustion of combining the amount of hydrogen with oxygen necessary to make the quantity of water in the oceans is just about right to produce a molten layer of rock 100 miles deep around the Earth. The heat of combustion of hydrogen and oxygen is 60,000 BTU/ lb and the volume of water in the ocean is 325 million cubic miles. The remainder of the molten core is the product of oxidation of the metals and possibly carbon. Chlorine/metal combustion obviously was involved in this combustion process, also. It is interesting to visualize the forming planetesimals igniting randomly throughout the dark of the Solar Nebula like fire flies.

Before planets were heated by the Sun, their atmospheric gases were solid nitrogen, CO_2 and water ice. Some hydrogen, helium and methane may also have been present. At that time, there was no significant oxygen left in the atmospheres since it had been consumed by combustion.

This is the point to discuss the T-tauri effect. The standard model has the solar winds sweeping off any atmospheres of the planets and then has volcanoes generating the atmospheres. The chemical reaction theory provides for a much more reasonable atmosphere than do volcanoes. The planets were far from the center and not yet in orbit when the Sun fired up. The atmospheric gases were layers of ice on the planets' surfaces and unaffected by the solar winds. Hydrogen and Helium would have been the only gases affected by the T-tauri. The ices liquefied and gasified only when they came sufficiently close to the Sun, much later than the T-tauri.

13. SOURCE OF CO$_2$.

The contribution of carbon in all of this planet formation is difficult to pin down. Most of it ended up as CO_2, but could have reached that structure by at least three different paths. First, it could have combined with oxygen ions before the oxygen became molecules. Second, the carbon could have become intermixed with the metal clouds and burned to form CO_2 while metal oxides were formed. Third, the carbon, like the metals, could have been in a carbon dust cloud that collapsed along with the gases, burned with the oxygen, and the entire mixture became gaseous. Regardless of the genesis, vast quantities of CO_2 were formed along with H_2O and these along with the nitrogen, hydrogen and helium collected around the planetesimals.

14.   ASSEMBLING THE SUN.

This is the easy part. Planets and planetesimals formed first and then along with the gases in their atmospheres and any remaining dust collapsed to become the Sun, slowly at first and then faster near the end. The Sun began its assembly near the center of the nebula. As matter began to coalesce near the center of the nebula, the forming Sun started pulling in all of the mass above, below and nearby in the ecliptic plane All mass, no matter how large, above and below the forming Sun would eventually be pulled in to the Sun. The gravitational pull on the Sun by masses around the Sun would tend to cancel each other and the Sun would remain in place and continue to grow. This explains the Sun's ending up with over 99.9% of the mass in the Solar System. In a nebula that has its mass evenly distributed, there is no net gravitational force on any material in the center, just as a person would not weigh anything in the center of the earth. The gravitational force on masses at the nebula level only becomes significant away from the center. For that reason any angular velocity of material such as gases, dust, debris, planetesimals and planets that are initially drawn to the center, will cause the material to veer to the right and travel past and away from the center until gravity pulls them back again. Such motion should result in continuous, chaotic merging and collisions until the mass at the center becomes great enough to trap the incoming material directly. The chaotic collisions during formation probably explain a big part of the fact that the

Sun has less angular momentum than its share of the Solar System's mass warrants. In fact, it is surprising the Sun ended up with any angular momentum at all. I have no feel for the timing of when the nuclear fires ignited on the Sun, but it was after the planets formed.

15. EARTH'S OXYGEN SOURCE AND CARBON MASS.

This subject is completely out of my area of expertise, but I thought I would include it because this is where I got started on this Solar System adventure. I was working on global warming and wondered where our Oxygen came from. That led to CO_2 and that led to planet building and away I went.

The previous description of the formation of the Earth described the combustion events that generated the water vapor, N_2 and CO_2 in the atmosphere of the early Earth. No significant amount of unburned oxygen remained in the atmosphere at that time. Sometime after the earth spiraled in and reached its orbit around the Sun and the Moon was in its place, conditions on the Earth became somewhat stable. The Sun and greenhouse effect melted the water, N_2, and CO_2 ices, oceans formed, and the Earth became a warm place. Large amounts of water vapor, hydrogen, helium, nitrogen, and CO_2 arrived with the planets as described above. The water vapor rained out and the hydrogen and helium were drawn off by the Sun. The planets were left with atmospheres that were predominately N_2, and CO_2. This conclusion can be estimated for earth from the quantities of nitrogen, carbonates and of free oxygen in the atmosphere today. The earth melted, the water in the oceans would have been highly acidic due to the high concentration of CO_2 in the atmosphere. The ocean would have been one big vat of soda water. Calcium hydroxide and other chemicals probably converted much of the original CO_2 to carbonates, but the oxygen that was in the atmosphere had to have been generated by an additional process.

This process began when an earth-changing event occurred. You can attribute it to lightning, virus mutation, chance, material from space, or God, but it happened. One small algae-like plant, scum, appeared in the sea that could live by photosynthesis and multiply itself. There could be no hard shelled life in the ocean at this time, due to the high acidity of the ocean. Since this little mother of us all had no competitors, no predators, and no other life forms to contend with, it spread like wildfire around the planet. This would have occurred over a really short period in geologic time since the scum had an almost infinite supply of CO_2 to consume in the atmosphere above. Oxygen molecules were freed up by the photosynthesis process as the algae spread. As new algae covered over the earlier algae, the older died from lack of sunlight and built layer upon layer. The dead material could not break down by the carbon cycle process because there were no microbes or plant eaters yet. The hydrogen and oxygen in the dead material would eventually be released and return to the sea as water, but the carbon bound together and stayed behind under water. This process continued for a very long time until the CO_2 in the atmosphere was nearly all consumed (0.038% is left in our air now). The amount of algae that had grown and piled up is a staggering number. We are talking 350 trillion tons of carbon. This would be a cube of coal 37 miles on a side. This amount of carbon would cover the entire earth with a one ft. depth of coal. This carbon rather than tropical plants and dinosaurs probably is the source of much of the coal and oil deep in the earth's crust. The algae based coal was formed millions of years before life on land. Plate tectonics would have buried much of that carbon. The conclusions above are based on the assumption that there were no other plausible sources for free oxygen on earth besides CO_2. There could be other sources, but CO_2 is an obvious candidate. Sea life came later and piled up shells and coral on top of the coal along with sand. It is interesting to note that the amount of carbon in the CO_2 and the carbonates in the earth's surface is about the same as the amount of carbon in Venus' atmosphere. Venus just never had any life to convert the CO_2. Mars should have water and CO_2 since it was formed by the same combustion process as Earth. The outer planets held on to their hydrogen and helium since they would have orbited far out and the Sun would not have captured these gases.

16. CONCLUDING REMARKS.

Developing this new theory has been a long and generally enjoyable experience. It has caused many sleepless nights wandering around the Solar System trying to figure out how things happened. If by some chance, observational data proves my theory wrong, that is the way of science. If that happens, if I were the builder, I would still build it my way.

ABOUT THE AUTHOR

James L. Baker is a retired Aerospace Engineer and lives in Huntsville, Alabama. He holds a BS in Aeronautical Engineering from the University of Kansas. Most of his 40 year career was on the manned space programs from Apollo through Shuttle. His specialty was fluid flow systems design and analysis. Most of his efforts involved work on liquid propellant rocket engines. Cryogenics was his area of expertise.

After retirement in 1996 James became an active environmentalist and gave presentations on alternative fuels and on the futility of sequestering carbon dioxide from coal-fired power plants. His interest in carbon dioxide is what led him to write this book on how the Solar System was built.

www.ingramcontent.com/pod-product-compliance
Lightning Source LLC
Chambersburg PA
CBHW041146180526
45159CB00002BB/740